THE TOP SECRET LIFE OF PLANTS

HOW PLANTS SPREAD SEEDS

BY JANEY LEVY

Gareth Stevens PUBLISHING

Please visit our website, www.garethstevens.com. For a free color catalog of all our high-quality books, call toll free 1-800-542-2595 or fax 1-877-542-2596.

Library of Congress Cataloging-in-Publication Data

Names: Levy, Janey, author.
Title: How plants spread seeds / Janey Levy.
Description: New York : Gareth Stevens Publishing, [2020] | Series: The top secret life of plants | Includes index.
Identifiers: LCCN 2018028865| ISBN 9781538233856 (library bound) | ISBN 9781538233832 (paperback) | ISBN 9781538233849 (6 pack)
Subjects: LCSH: Seeds–Dispersal–Juvenile literature.
Classification: LCC QK929 .L48 2020 | DDC 581.4/67–dc23
LC record available at https://lccn.loc.gov/2018028865

First Edition

Published in 2020 by
Gareth Stevens Publishing
111 East 14th Street, Suite 349
New York, NY 10003

Designer: Sarah Liddell
Editor: Abby Badach Doyle

Photo credits: Cover, p. 1 Przemek Klos/Shutterstock.com; glass dome shape used throughout bombybamby/Shutterstock.com; leaves used throughout janniwet/Shutterstock.com; background texture used throughout MInerva Studio/Shutterstock.com; p. 5 hacohob/Shutterstock.com; p. 7 Paula French/EyeEm/EyeEm/Getty Images; p. 9 (mangrove trees) benedix/Shutterstock.com; p. 9 (dandelion) Samuel Borges Photography/Shutterstock.com; p. 11 Bernard/Getty Images; p. 13 Vortbot/Wikimedia Commons; p. 15 Katja Pruess/Shutterstock.com; p. 17 jurra8/Shutterstock.com; p. 19 (wind) Bess Hamitii/Shutterstock.com; p. 19 (water) keisuke/Shutterstock.com; p. 19 (explosive force) Artistas/Shutterstock.com; p. 19 (eaten) (c) Niranj Vaidyanathan v.niranj@gmail.com/Moment/Getty Images; p. 19 (hitchhiking) Jackan/Shutterstock.com; p. 19 (buried) Alex Zabusik/Shutterstock.com; p. 19 (fire) Lumppini/Shutterstock.com; p. 21 Designua/Shutterstock.com.

Printed in the United States of America

CPSIA compliance information: Batch #CS19GS: For further information contact Gareth Stevens, New York, New York at 1-800-542-2595.

CONTENTS

Words in the glossary appear in **bold** type the first time they are used in the text.

WHY SEEDS MUST LEAVE HOME

Seeds are plants' children. Plants want their seeds to fall on good ground, where they'll be able to **germinate** and grow. That means the seeds must land where they have enough **nutrients**, water, and sunlight. If all the seeds just drop beneath the parent plant, they'll be overcrowded. They'll fight for what they need, and many won't survive. So seeds must be spread far away.

Plants have invented some pretty creative ways of spreading seeds. What are their secrets? Keep reading to learn the story of seed **dispersal**.

CLASSIFIED!

IF ALL THE SEEDS FALL AND GROW IN ONE PLACE, THEY'RE IN DANGER IF FIRE OR SICKNESS BREAKS OUT. IF THEY'RE SPREAD FAR APART, THERE'S LESS DANGER THAT ALL WILL BE HARMED.

PLANTS HAVE MANY DIFFERENT KINDS OF SEEDS. SOME MAY BE HELD INSIDE ANOTHER PLANT PART, SUCH AS FRUIT OR A PINECONE.

HOW SEEDS ARE BORN

Before we look at dispersal, it's good to know more about the seeds themselves. Beneath a seed's tough shell is a secret food supply to help the baby plant get started. That's built into how seeds are formed.

Seeds are formed through pollination, or when pollen is carried from the male part of one flower to the female part of another. There, the pollen makes a seed. It includes the baby plant and the food supply. Bugs, hummingbirds, bats, or wind help the pollen move.

CLASSIFIED!

THE BABY PLANT IS CALLED THE EMBRYO (*EHM*-BREE-OH). THE FOOD SUPPLY IS CALLED THE ENDOSPERM (*EHN*-DOH-SPERM).

THIS BEE IS COVERED WITH POLLEN, WHICH IT WILL CARRY TO THE NEXT FLOWER IT VISITS.

WIND AND WATER

Have you ever blown on a **dandelion** to watch its light, feathery seeds float away on the wind? It's among many plants that use wind for seed dispersal. Some of these plants have seeds much like those of the dandelion. Other plants, like the maple tree, have seeds that seem to have wings.

Another method is water dispersal. Alder trees and mangrove trees grow near water and depend on it to spread their seeds. The seeds float away to new places where they sprout and grow near water, too.

DANDELIONS GROW THROUGHOUT MUCH OF THE WORLD. MANGROVE TREES LIVE IN WARM AREAS ALONG COASTS.

mangrove trees

CLASSIFIED!

SEEDS CARRIED BY WIND MAY FALL IN PLACES WHERE THEY CAN'T GROW, SO PLANTS THAT DEPEND ON WIND DISPERSAL USUALLY MAKE LOTS OF SEEDS.

dandelion

EXPLOSIVE FORCE

It might be hard to believe, but some plants' secret to dispersal is an **explosive** force. One such plant is the pea plant. The peas, which are the plant's seeds, grow in pods. The drying pods twist and break open, popping out the peas.

CLASSIFIED!

THE SEEDS OF THE SANDBOX TREE CAN TRAVEL AT SPEEDS UP TO 160 MILES (257 KM) PER HOUR!

An African tree in the same family as peas can shoot its seeds more than 150 feet (46 m). The sandbox tree, which lives in hot areas of the Americas, can shoot its seeds up to 330 feet (100 m)!

PEOPLE HAVE BEEN HURT BY THE FORCE OF THE SEEDS EXPLODING FROM THE SANDBOX TREE.

EATEN AND DIGESTED

Some plants' secret to dispersal is getting birds and other animals to eat their fruit that holds the seeds. The birds and other animals then drop the seeds far from the parent plant. Or they eat the seeds, the seeds pass through their **digestive system** unharmed, and then come out in their poop.

However, some seeds need to be **digested**! The hard seeds of cherries and blackberries need the digestive system to break them down in order for them to germinate.

CLASSIFIED!

ALMOST THREE-QUARTERS OF THE TREES IN NEW ZEALAND FORESTS PRODUCE FRUIT WITH SEEDS THAT MUST BE DIGESTED BY NATIVE BIRDS.

THE KERERU, A TYPE OF PIGEON, IS AN IMPORTANT BIRD FOR SPREADING SEEDS IN NEW ZEALAND. ITS DIGESTIVE SYSTEM HELPS WEAKEN THE HARD COAT ON SEEDS.

HITCHHIKING

Some plants have a sneaky secret for spreading their seeds. Their seeds grab a ride—uninvited—on passing animals, including people! The seeds have **hooks** or sticky stuff on them, and they stick to birds, other animals, and people that brush against them.

The birds, other animals, and people will get rid of their uninvited guests later on. But by that time, the seeds are usually far from the parent plant. Twinflower and cocklebur are two plants that spread seeds this way.

CLASSIFIED!

SEEDS THAT SPREAD THIS WAY DON'T USUALLY CAUSE HARM TO THE CREATURES THAT SPREAD THEM. BUT FOXTAIL SEEDS CAN GET STUCK IN AN ANIMAL'S EARS AND NOSE AND HURT IT.

LOOK CLOSELY AT THIS COCKLEBUR SEED. CAN YOU SEE THE LITTLE HOOKS IT USES TO GRAB A RIDE ON AN ANIMAL'S FUR?

BURIED ALIVE!

One dispersal secret used by some plants actually sounds pretty awful. The seeds are buried alive! But don't worry. That's exactly what's supposed to happen.

Acorns are the heavy seeds of oak trees. Squirrels helpfully carry them off and bury them to eat during winter. But squirrels often forget where some of their acorns are buried, and those acorns germinate and grow. In Europe, jays rather than squirrels bury acorns. They may bury up to 4,600 in one autumn!

CLASSIFIED!

IN EUROPE, RED SQUIRRELS AND MICE CARRY OFF AND HIDE **HAZELNUTS** TO EAT LATER. THE ONES THEY DON'T EAT GERMINATE AND GROW.

AN ACORN OR NUT PARTLY EATEN BY A SQUIRREL CAN STILL GROW AS LONG AS THE BABY PLANT INSIDE HASN'T BEEN HARMED.

FREED BY FIRE

Unlike animals, plants can't run away from a fire. We often think of fire as something that is harmful or even destroys things. But for some plants, fire is a good thing!

Fire can help a plant spread seeds. For example, some kinds of pine trees need a fire's heat to help their pinecones open and let seeds out. Fires are common in some places, like Australia and the western United States. Plants that grow in those places have learned to adapt to fires.

CLASSIFIED!

A FIRE HAS TO BE JUST RIGHT FOR SEED DISPERSAL TO WORK. PLANTS WON'T GROW BACK IF A FIRE IS TOO BIG, TOO HOT, OR BURNS TOO LONG.

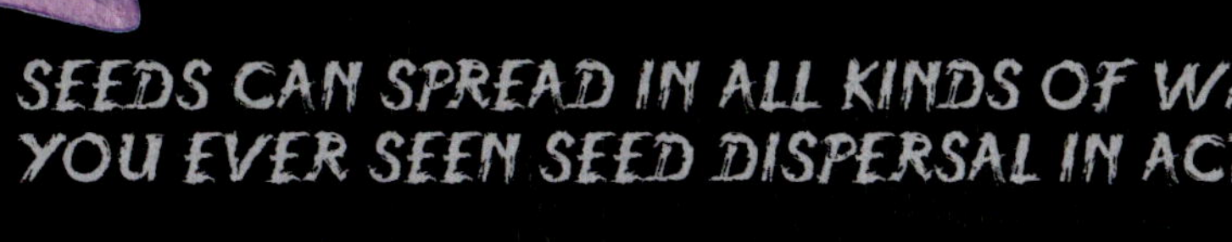

SEEDS CAN SPREAD IN ALL KINDS OF WAYS. HAVE YOU EVER SEEN SEED DISPERSAL IN ACTION?

TYPES OF SEED DISPERSAL

WIND

WATER

EXPLOSIVE FORCE

EATEN

HITCHHIKING

BURIED

FIRE

SEEDS BECOME PLANTS

The seeds that have left home will slowly become new plants. They will send roots down into the soil to gather water and nutrients. A stem will grow up above the ground and produce leaves. The leaves will make food for the plant through **photosynthesis** (foh-toh-SIHN-thuh-suhs).

In time, the plant will make flowers. Then the plant's **life cycle** starts all over again. You know the rest from here: Pollination occurs, new seeds grow, and the seeds are sent out into the world!

CLASSIFIED!

NOT ALL THE SEEDS WILL GROW. ONLY THE SEEDS THAT LAND WHERE THEY HAVE ENOUGH NUTRIENTS, WATER, AND SUNLIGHT WILL BE ABLE TO GERMINATE AND GROW.

A PLANT'S LIFE CYCLE CREATES SEEDS THAT TRAVEL AND GROW INTO NEW, HEALTHY PLANTS.

LIFE CYCLE OF AN APPLE TREE

FLOWER

FRUIT

SEEDS

YOUNG PLANT

ADULT PLANT

GLOSSARY

adapt: to change to suit conditions

dandelion: a common wild plant with a bright yellow flower that is later followed with a ball of white, light, feathery seeds

digest: to break down food inside the body so that the body can use it

digestive system: all the body parts concerned with eating, breaking down, and taking in food

dispersal: the act or result of spreading around

explosive: happening suddenly or quickly

germinate: to begin to grow

hazelnut: the seed of the hazel tree

hook: a piece of hard matter bent into a curve

life cycle: the stages through which a living thing passes in its life

nutrient: something a living thing needs to grow and stay alive

photosynthesis: the method by which a plant makes food using water, nutrients, sunlight, and a gas called carbon dioxide from the air

FOR MORE INFORMATION

BOOKS

Aston, Dianna Hutts. *A Seed Is Sleepy*. San Francisco, CA: Chronicle Books, 2014.

Macken, JoAnn Early. *Flip, Float, Fly: Seeds on the Move*. New York, NY: Holiday House, 2016.

Owen, Ruth. *How Do Plants Make and Spread Their Seeds?* New York, NY: PowerKids Press, 2015.

WEBSITES

BBC: Seed Dispersal
www.bbc.co.uk/nature/adaptations/Seed_dispersal#intro
Watch some videos about different ways seeds spread here.

PBS Kids: Seed Racer
pbskids.org/plumlanding/games/seed_racer/
Move fast to collect seeds on the move in this interactive game.

Seeds on the Move–Seed Dispersal for Kids
www.kidsdiscover.com/parentresources/seed-dispersal/
Learn more about how plants spread seeds on this website.

INDEX